A Chemotherapy Guide

for the Medical-Surgical Nurse

A Chemotherapy Guide

for the Medical-Surgical Nurse

By

Carol L. Ackerman MSN, RN

Copyright Year: 2007

Copyright Notice: © by Carol L. Ackerman. All rights reserved.

License: This work is licensed under the Creative Commons Attribution-NonCommerical-ShareAlike License. To view a copy of this license, visit http://creativecommons.org/licenses or send a letter to Creative Commons, 543 Howard Street, 5th Floor, San Francisco, California, 94105, USA.

Published and distributed by:
Lulu, Inc.
860 Aviation Parkway
Suite 300
Morrisville, N.C. 27560

ISBN 978 – 1– 4303 – 2870 - 4

This book is dedicated to my sisters, Linda Penn and Irene Alveraz, who have been cancer survivors since 1992 and to the memory of my brother-in-law, Thomas Penn, who lost his long and courageous battle against Non-Hodgkin's Lymphoma in 2003.

ABOUT THE AUTHOR

Carol L. Ackerman is a registered nurse presently employed as a charge nurse (PNCC) on a 24 bed medical-surgical unit with an oncology focus. She received her basic nursing education from Community College of Allegheny County, West Mifflin, Pennsylvania in 1979. She was employed as a staff nurse at UPMC Braddock from 1979 to 2004. In 2003, she returned to college and earned her BSN from California University of Pennsylvania (CUP) in 2005 and her MSN from Indiana University of Pennsylvania (IUP) in 2007. It was during her practicum experience for IUP that she developed these guidelines for the medical-surgical nurse to provide safe care for a patient receiving chemotherapy. Carol is also the proud mother of Bethany and grandmother of Lucas Gregg.

TABLE OF CONTENTS

Pre Test..1

Introduction...3

Objectives...4

Chemotherapy Drugs..5

Exposure...7

Personal Protective Equipment...8

Waste Containers..10

Signs...11

Spill Kits...13

Spill Management..15

IV Complications...16

IV Monitoring..21

Patient Education...23

Definitions...30

Attachments..31

Post Test..39

Answers to Pre/Post Tests...41

Acknowledgements...42

References...43

Chemotherapy Precautions
Test Your Knowledge
Pre-test

Directions: Select the correct answer for each of the following questions.

1. Using the safety guidelines for chemotherapy administration, what is the designated chemotherapy excretion safety period?

   _____48 hours for all drugs
   _____72 hours for all drugs
   _____7 days for all drugs
   _____72 hours or 7 days, depending on the drug

2. What agency has established guidelines for the handling and administration of chemotherapy agents for the health care professional?

   _____Joint Commission on Accreditation of Healthcare Organizations (JACHO)
   _____American Cancer Society (ACS)
   _____National Cancer Institute (NCI)
   _____Occupational Safety and Health Administration (OSHA)

3. When handling the waste of individuals receiving chemotherapy, adequate protection includes

   _____gloves.
   _____gloves, gown, face protection.
   _____gloves, gown.
   _____gloves, face protection.

4. Wearing two pairs of chemotherapy gloves when a gown is worn serves to protect the skin from contamination while the gown is being removed. Listed below are the steps to be followed when applying the two pairs of chemotherapy gloves and the gown. Number the steps from 1 to 3 to correspond with the correct order in which you would perform these activities:

   _____apply the second pair of gloves
   _____apply the first pair of gloves
   _____apply the gown

5. When transporting a patient who is receiving chemotherapy to the radiology department, the intravenous catheter becomes dislodged and the chemotherapy agent spills onto the floor. What action would you take?

_____Take the patient back to his room and call housekeeping to clean the spill
_____Secure the area so no one is accidentally exposed before the spill can be cleaned up and follow the directions on the spill kit for cleanup
_____Reinsert the intravenous catheter, call housekeeping to clean up the spill, and continue with the patient to the radiology department
_____Call the nurse's aide assigned to that team to clean up the spill

6. The contents of the Chemotherapy Drug Spill Kit includes all but

_____shoe covering.
_____surgical mask.
_____chemo waste bags.
_____a scooper with detachable scraper.

7. Your patient has developed small mouth sores after receiving his chemotherapy. Teaching should include

_____do not drink through a straw as the straw puts pressure on the ulcers causing more pain.
_____any type of over-the-counter mouthwash will be effective in reducing the sores.
_____avoid hard candy or gum if your mouth or throat is sore.
_____avoid citrus fruits, tomatoes, and alcoholic beverages.

True or False:

8. Yellow waste bags are left in the patient's room when a patient is receiving chemotherapy until either the chemotherapy is finished or the bag becomes full.

_____True   _____False

9. A chemotherapy trained registered nurse must monitor the chemotherapy infusion.

_____True   _____False

10. A significant chemotherapy spill is 5 milliliters or larger.

_____True   _____False

INTRODUCTION

Hazardous drugs are used in the treatment of persons with cancer. They can also be used for non-oncology indications such as rheumatoid arthritis, lupus, nephritis, and multiple sclerosis (Polovich, 2004). These agents are referred to as cytotoxic agents, chemotherapy agents, or antineoplastic agents (Attachment "A"). These agents are toxic to all living cells, but are selective against rapidly growing cancer cells.

Chemotherapeutic agents have by-products that are excreted in the urine and feces. This poses a hazard for healthcare workers. Healthcare professionals who are administering or taking care of patients receiving chemotherapy must be protected against exposure when handling any amount of these drugs. The United States Department of Labor, Occupational Safety and Health Administration (OSHA) have developed guidelines for the safe handling of chemotherapy agents (Polovich, 2004).

Most chemotherapy agents are administered in specialty areas such as cancer centers or oncology units in the hospital; however, this is not necessarily the case. With our rapidly changing health care systems, some hospitals do not have oncology units. Some units are classified as a medical-surgical unit with oncology overflow or focus. They are staffed with medical-surgical nurses. Only a small percentage of the nurses assigned to that unit are chemotherapy trained or certified. On these units, the chemotherapy agent is administered by a Chemotherapy Certified Nurse and monitored by the staff nurse.

The purpose of this booklet is to provide the staff nurse with guidelines for keeping their patients safe during chemotherapy administration.

OBJECTIVES

Upon completion, the registered nurse will:

1. State the four ways that healthcare workers can become exposed to hazardous drugs.
2. Demonstrate the correct application of the personal protective equipment.
3. Verbalize the proper procedure for cleaning up a chemotherapy spill.
4. Explain the different intravenous complications.
5. Successfully complete post test by obtaining 100% accuracy.

CHEMOTHERAPY DRUGS

Chemotherapy drugs are used to treat cancer. Other medications are used to treat the side effects of the chemotherapy drugs. Chemotherapy medications can be given orally, topically, intravenously, intrathecally, or by injection. The list of chemotherapy drugs is not conclusive because drugs are occasionally added to or deleted from the list (Attachment "A").

Chemotherapy drugs are classified as to how they perform. There are five different classifications of chemotherapy drugs.

ALKYLATING: This is the oldest classification of chemotherapeutic drugs. These drugs work in all phases of the cell cycle. They directly attack the DNA, which stops the activity of the cell, effectively killing it.

ANTIMETABOLITES: These drugs interfere with specific metabolic pathways and block enzymes necessary for the DNA and RNA of the cancer cells to grow. These drugs attack during the cell division process. These drugs imitate normal cell nutrients and fool the cancer cells into consuming them.

ANTITUMOR ANTIBIOTICS: These drugs interfere with cancer cell DNA by blocking certain enzymes and cell division (mitosis). They charge the membranes that surround the cells and either break up chromosomes or retard the synthesis of the RNA that the cell needs to grow. These drugs work in all phases of the cell cycle. These antibiotics are not the same antibiotics that are used to treat infections.

MITOTIC INHIBITORS: These drugs are plant alkaloids and other compounds acquired from natural products. These drugs stop cell division and stop enzymes from making the proteins that are required for cell production.

STEROID HORMONES: Steroids are naturally produced in the body. Some steroids slow the growth of cancer cells. Other steroids destroy the cancer cells, making chemotherapy more effective (American Cancer Society, 2006).

EXPOSURE

Exposure to chemotherapeutic agents can occur when safe handling methods fail or when they are not used. Exposure may occur during preparation, transport, administration, when handling patient excretions, during disposal of chemotherapy agents, or in the event of a spill. There are four ways in which health care workers can be exposed to hazardous drugs:

INHALATION Exposure occurs when you breathe the chemotherapy agent into your lungs and into your blood stream. Aerosols can be present when removing contaminated items, such as gloves, syringes, or vials.

ABSORPTION Exposure occurs when the chemotherapy agent comes into contact with your skin and is absorbed into the blood stream. Absorption can also occur through your mucous membranes, or eyes, from contact with contaminated skin.

INGESTION Drug ingestion occurs as a result of direct drug contact with food, beverages, chewing gum, food containers, utensils, or tobacco products. Ingestion may also occur when a contaminated hand comes in contact with the mouth. Small amounts of the chemotherapy agent are swallowed.

INJECTION Accidental needle stick injury from sharp objects. It is recommended that you do not recap your needles.

Care must be taken when handling or administering chemotherapy to reduce the health care worker's risk of exposure.

PERSONAL PROTECTIVE EQUIPMENT

Personal Protective Equipment (PPE) should be consistently used when handling chemotherapeutic agents. Failure to use the appropriate equipment may result in exposure. Personal protective equipment is recommended by OSHA to reduce your chances of occupational exposure.

GOWNS: should be disposable and made of a low permeable fabric. They should have a closed front, long sleeves, and elastic or knit cuffs (OSHA, 2006).

GLOVES: should be powder free. They should be made of a non-latex product such as nitril, neoprene, or natural rubber.

FACE SHIELDS OR GOGGLES: should be used for protection if a splash is possible. Face shields cover the entire face and provides more protection than goggles.

RESPIRATORY MASKS: are used when cleaning up spills. A fit test N-95 respirator mask should be worn. A surgical mask is not a respirator and does not protect against aerosols or vapors (Polovich, 2004).

HOW TO PROPERLY APPLY PPE

When administering chemotherapy or caring for the patient receiving chemotherapy, you should apply a disposal gown and two pair of gloves. The correct way to apply the PPE:

- Put on the first pair of gloves. (Picture 1)
- Put on the disposal gown, fastening it closed. Make sure the gloves are <u>under</u> the cuffs of the gown. (Picture 2)
- Apply the second pair of gloves, making sure they are <u>over</u> the cuffs of the gown. (Picture 3)
- You may wear goggles or face shield to protect yourself in the event that a spill or splash may occur.

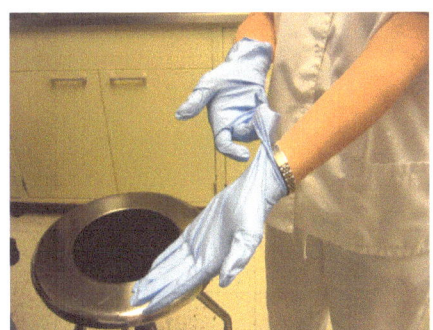

Picture 1

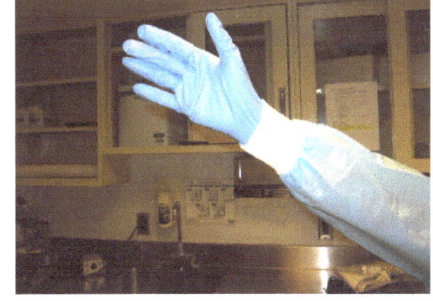

Picture 2

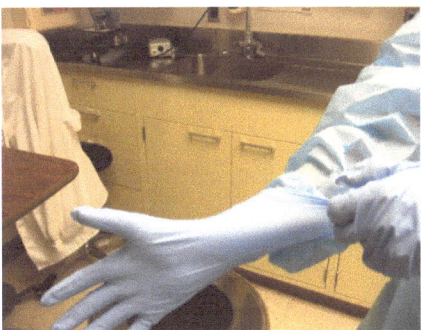

Picture 3

WASTE CONTAINERS

Yellow waste containers are used for the disposal of chemotherapy waste. These containers come in a variety of sizes. Disposal of these containers should be in accordance with OSHA guidelines.

- Yellow bag: Used to dispose of soft bulky items such as gowns, gloves, masks, and absorbent pads. Yellow bags should be disposed of daily. Do not wait until the yellow bags are full or the chemotherapy is completed.

- Large yellow bin: Placed in the patient's room who is either receiving chemotherapy or who is maintained on chemotherapy precautions. Used for disposal of gowns, gloves, or masks after patient care or contact.

- Small rigid yellow container: Used for the disposal of sharps, bags, bottles, and administration sets. Usually used if chemotherapy is being given short term. May also be used for the disposal of oral agents. Oral agents require administration with gloves. The pill package and gloves must be disposed of in a yellow waste container.

- Large rigid yellow container: Used if chemotherapy is being given for a longer term. This container is also used for the disposal of sharps, syringes, tubing, or intravenous bags.

CHEMOTHERAPY SIGN

There are two signs used to identify patients who may be receiving chemotherapy or when chemotherapy precautions are being maintained.

Chemotherapy Excretion Safety: This yellow sign is used whenever the patient is receiving chemotherapy or is in chemotherapy precautions. One sign is placed outside the patient's room and another is placed in the patient's bathroom.

> **CHEMOTHERAPY EXCRETION SAFETY**
>
> When handling patient's urine, feces, emesis, sputum, or other body fluids for _____ 3 days or _____ 7 days after chemotherapy, family and health team must:
>
> 1. Wear Personal Protective Equipment.
>   - One or Two pairs of approved chemotherapy gloves (two pairs of gloves must be worn if a gown is worn)
>   - An approved chemotherapy gown whenever potential exposure of the body to soiled linen, urine, or feces is anticipated
>   - A face shield if splashing is likely
>
> 2. Flush toilet immediately after use.
>
> 3. Dispose of gloves, and other soiled disposable materials in the chemotherapy waste container.
>
> 4. Wash hands after removing gloves.
>
> <u>**INDIVIDUALS WHO ARE PREGNANT, LACTATING, OR TRYING TO CONCEIVE SHOULD AVOID UNPROTECTED CONTACT WITH BODY FLUIDS DURING THIS TIME PERIOD.**</u>
>
> Start date / time: _____   Finish date / time: _____
>
> FORM # 4994-71230-0106

Chemotherapy precautions continue for 72 hours or up to seven days after the chemotherapy has been discontinued. The chemotherapy usually remains in the body for that period of time and precautions should be maintained.

NEUTROPENIC SIGN

Neutropenic Precautions: This red sign is utilized whenever the patient is neutropenic. When a patient is neutropenic, they have a low number of neutrophils in their blood. Since the neutrophils destroy bacteria in the blood, they are the body's defense against infections. Neutropenia can be life threatening for the patient who is receiving chemotherapy. This sign may be used in addition to the Chemotherapy Excretion Safety sign. This sign is placed outside the patient's door.

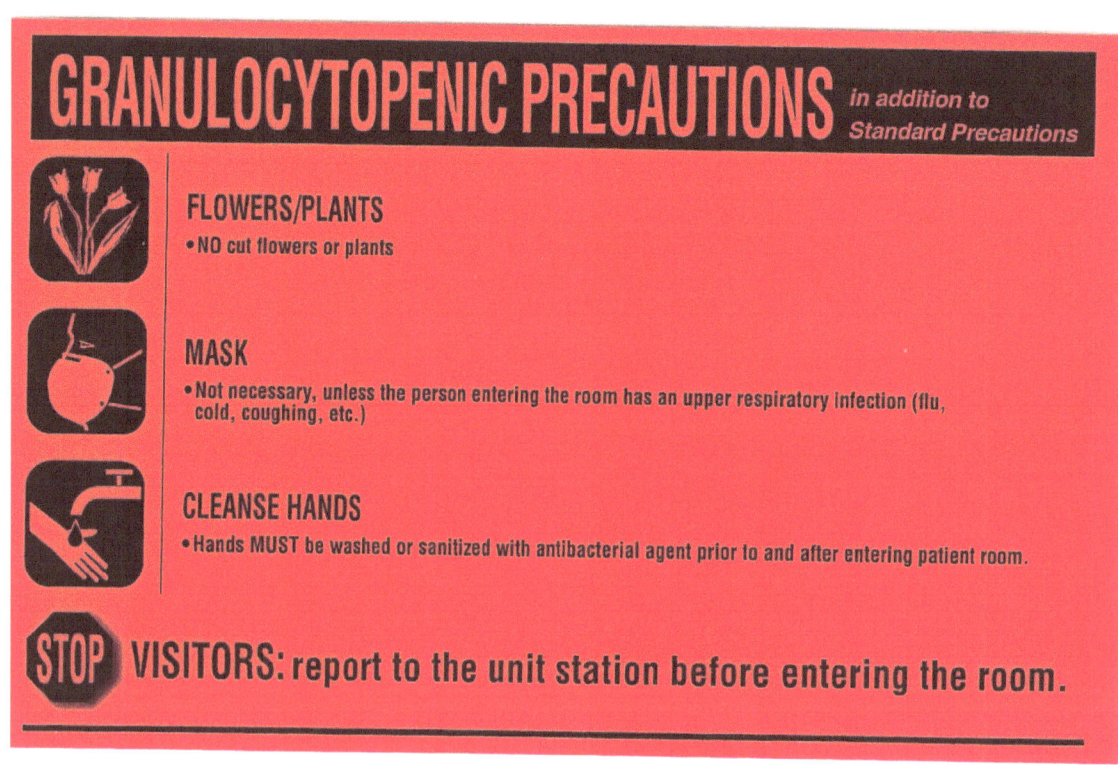

SPILL KITS

Even though extreme caution is taken when handling chemotherapeutic agents, some spills are unavoidable. A significant spill is 5 milliliters or larger (OSHA, 2006). An incident report is always filled out if a spill occurs. Spill kits should be available wherever chemotherapy drugs are stored, prepared, or administered. A spill kit should be kept in the patient's room during the administration of chemotherapy. The spill kit should also accompany the patient when the patient is transported to and from another department, if the patient is actively receiving chemotherapy.

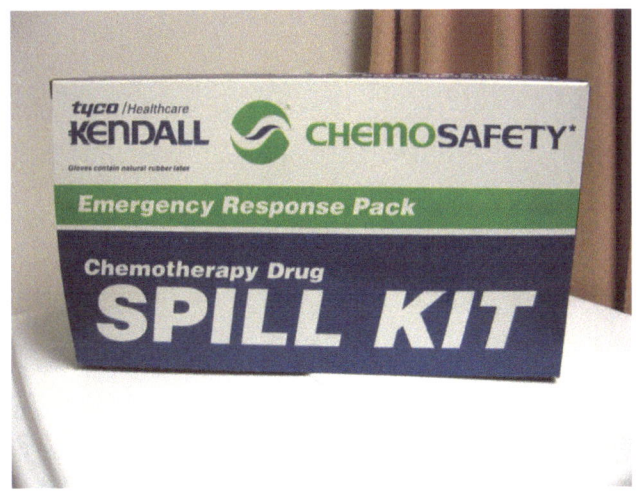

SPILL KIT CONTENTS

Contents of a spill kit:

2 pairs of chemo gloves	1 chemotherapy gown
1 pair of safety glasses	1 respirator mask
3 spill towels	2 chemo waste bags
1 pair shoe covers	1 sign
2 chemo absorbent pads	1 scoop with detachable scraper
1 tie wrap	1 green hazardous waste tag

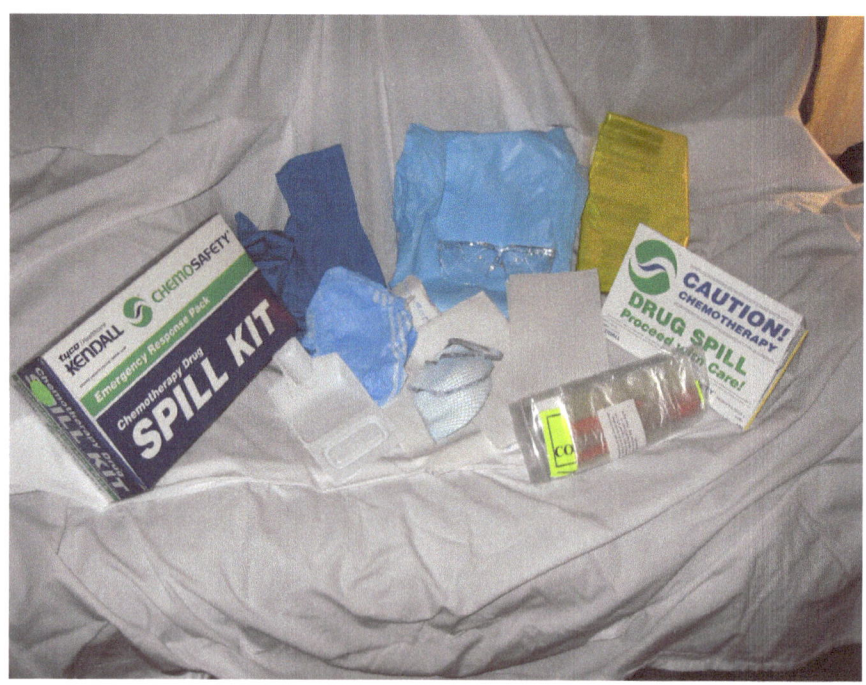

CHEMOTHERAPY SPILL MANAGEMENT

KEEP EVERYONE AWAY FROM SPILL SITE

1. Take out contents of kit. Display sign near spill area.

2. Put on chemo gloves, chemo gown, respirator mask, safety goggles, and 2nd pair of chemo gloves.

3. Lay absorbent pads over spill. The pads will absorb the liquid and transform it into a gel. The gel is extremely slippery when wet. Avoid skin and eye contact. Do not inhale.

4. Detach scoop from scraper and use both to pick up gel. Place contaminated gel in one of the waste bags. If there is any broken glass, use scoop to place it in yellow hard chemo container.

5. Use spill towels, detergent, and water to pick up any remaining gel. Place towels in the same waste bag.

6. Remove respirator mask, safety goggles, shoe coverings, gown and outer pair of gloves. Discard in same waste bag.

7. Wearing only the inner gloves, close the bag and place it in the second chemo waste bag. Remove gloves and discard in the chemo waste bag. Close bag with tie wrap. Place green hazardous waste tag on outer bag.

8. Dispose of bag according to facility regulations (Kendall, 2005).

IV COMPLICATIONS

Intravenous therapy or IV therapy is the giving of a liquid substance directly into a vein. This can be intermittent or continuous. Some of the complications that may result in intravenous therapy can range from a minor flare to a more severe extravasation. Other complications include infiltration, infection, and phlebitis.

Flare: a painless local allergic reaction. A blood return is usually present. It may be accompanied by a red streak, hives, or itchiness.

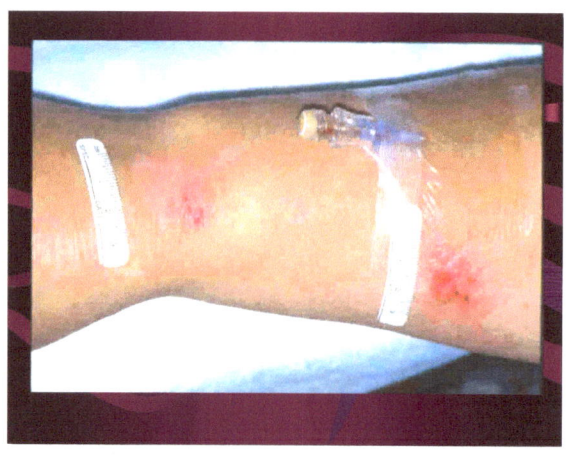

Photograph courtesy of Gloria Gotaskie, MSN, RN
Hillman Cancer Center,
Pittsburgh, Pennsylvania

Infiltration: the inadvertent administration of nonvesicant medication or fluid into the surrounding tissue instead of into the intended vascular pathway (Rosenthal, 2006). It occurs when the infusion cannula is no longer fully positioned in the vein.

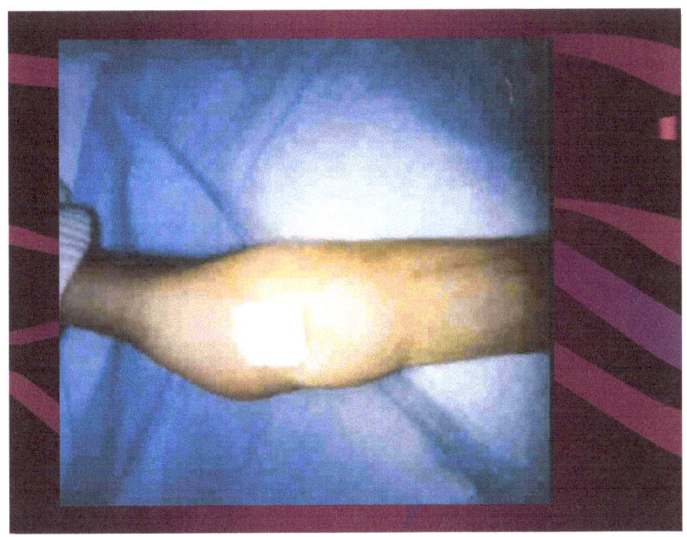

Photograph courtesy of Gloria Gotaskie, MSN, RN
Hillman Cancer Center,
Pittsburgh, Pennsylvania

Infection: results when aseptic techniques are not adhered to when intravenous line is inserted. Aseptic techniques are vital in the prevention of intravenous related infections. Asepsis should be maintained at insertion, during clinical use, and when removing the device.

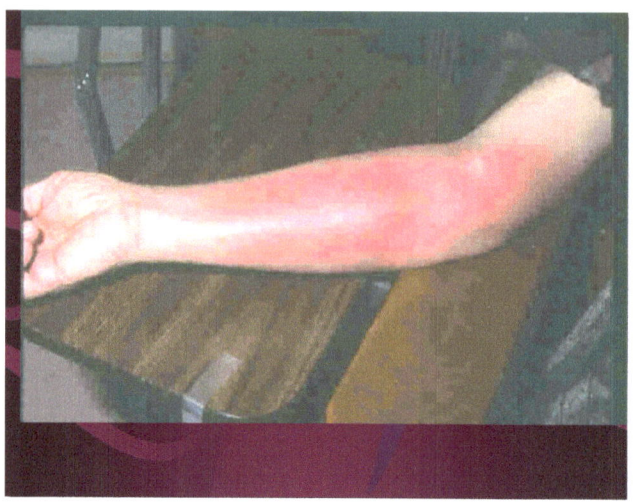

Photograph courtesy of Gloria Gotaskie, MSN, RN
Hillman Cancer Center,
Pittsburgh, Pennsylvania

Phlebitis: irritation of a vein that is not caused by infection, but from the mere presence of a foreign body (the IV catheter) or the fluids or medication being given. It is a condition where the veins close to the surface of the body (superficial) are inflamed. Symptoms are swelling, pain, and redness around the vein. It does not necessarily mean the IV device must be removed. Warmth, elevation of the affected limb, or a change in the rate of flow may resolve the symptoms. Due to frequent injections and recurring phlebitis, the peripheral veins of cancer patients undergoing chemotherapy become hardened and difficult to access over time (Rosenthal, 2006).

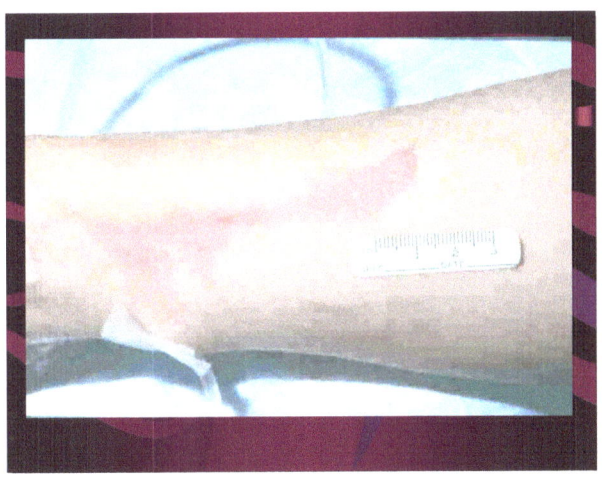

Photograph courtesy of Gloria Gotaskie, MSN, RN
Hillman Cancer Center,
Pittsburgh, Pennsylvania

Extravasation: refers to the leakage of a fluid out of its container. Extravasation is the accidental administration of intravenously (IV) infused medicinal drugs into the surrounding tissue, either by leakage (example: because of brittle veins in very elderly patients), or directly (example: because the needle has punctured the vein and the infusion goes directly into the arm tissue). Extravasation can cause: 1) pain, reddening, irritation, (slight damage) on the arm with the infusion needle, or 2) severe damage up to tissue necrosis (parts of the arm tissue die). It can lead to a loss of an arm in extreme cases (Rosenthal, 2006).

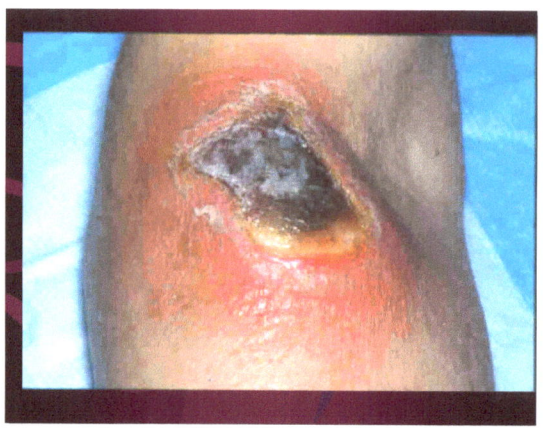

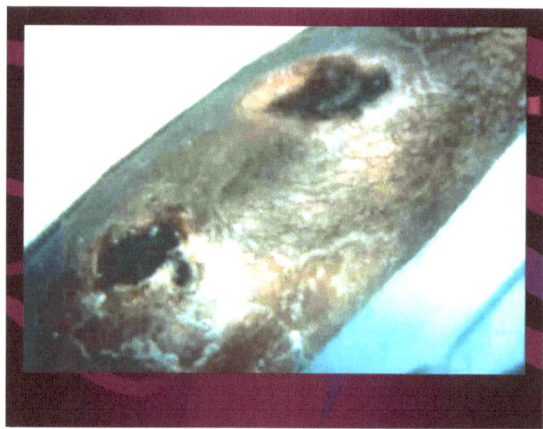

Photographs courtesy of Gloria Gotaskie, MSN, RN
Hillman Cancer Center,
Pittsburgh, Pennsylvania

IV MONITORING

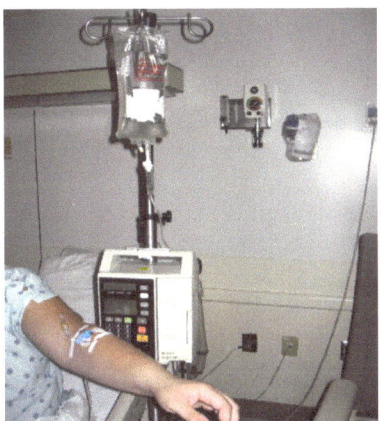

To avoid or minimize any of the before mentioned complications, it is imperative that the chemotherapeutic drugs are monitored continuously. The chemotherapy is administered by a chemotherapy trained registered nurse but can be monitored by a staff nurse on a medical-surgical unit.

Some guidelines include:

- Monitor the intravenous site hourly. Assess for redness, swelling, or tenderness at the site. Notify the chemotherapy nurse if any of these signs occur.

- Monitor the chemotherapy on a separate IV sheet. If another intravenous is infusing, use two intravenous sheets.

- Do not connect any other intravenous medications to the chemotherapy infusion.

- Every hour note the rate of infusion, amount of chemotherapy infused, and amount remaining in the infusion bag.

- Assess for blood return every four hours if patient is on a continuous intravenous infusion.
- Do not use a double infusion pump. If another intravenous is infusing, it is preferable to use the opposite limb.
- Attach a yellow chemotherapy sign to the pump that contains the infusing chemotherapy. This will alert all healthcare workers that the chemotherapy is infusing in that particular site.
- Do not adjust the rate or trouble shoot a pump that has chemotherapy infusing. Notify the chemotherapy nurse if any problems occur.
- Never disconnect an intravenous that has chemotherapy infusing. If the chemotherapy is finished, notify the chemotherapy nurse.
- Consider using the pump alarm system. This is a specially designed system that provides a distinct ring at the nurse's station to alert the staff that immediate attention should be given to the infusion.
- See ATTACHMENT "B" for sample intravenous monitoring sheet.

PATIENT EDUCATION

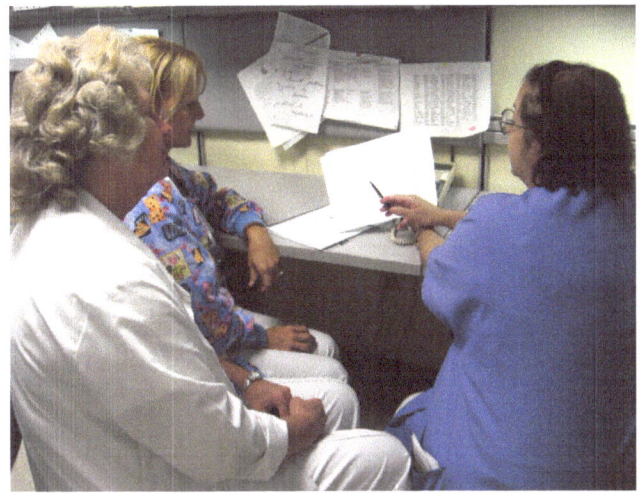

      Patient education is a planned learning experience. Patient education is a critical part of nursing care and has proven benefits to patients and their caregivers. Patients should be educated about their chemotherapy medications, side effects, and complications. The teaching should be documented in the patient's chart for ethical and legal reasons.

MEDICATIONS AND SIDE EFFECTS

Each patient should be aware of the medications they are receiving and the side effects that may occur as a result of receiving them. Patients should be educated on the name of the medication, reason for taking the medication, dosage, and how often it is administered. Patients should also be instructed on whether the medication should be taken on an empty stomach or with food. The side effects of the medication and how to manage the side effects should be included in patient education.

- Print copy of medication sheet and give to patient.
- Explain medication and side effects to patient.
- Have patient verbalize understanding of chemotherapy medication.
- Educate the patient about all of the medications being used to manage pain, nausea, vomiting, or diarrhea, as well as the chemotherapeutic agent being administered.

Information about medications can be obtained by accessing the internet for information. Some institutions have websites or educational channels that can be utilized for patient education.

PAIN

Pain affects over half of cancer patients at some time during the course of the disease. Cancer pain is the result of tumor growth or cancer treatment related side effects. Not all patients with cancer will experience pain.

- The nurse will assess your pain on a scale of zero to ten with ten being the most intense.
- Pain can also be described as mild, moderate, severe, or excruciating.
- The Wong-Baker scale may also be used in describing pain. Pictures of faces with smiles or frowns are utilized. The nurse will ask you to choose the face that best describes your pain.
- Do not wait until you are having severe pain to request medication. The longer you wait, the longer it will take for the medication to become effective.
- Take pain medication prior to any activity since pain increases during activity.
- The nurse will periodically ask you how bad your pain is at that given time.
- If the patient has unrelieved pain, the physician must be consulted for pain management.

LOSS OF APPETITE

This is a temporary side effect that may occur as a result of receiving chemotherapy.

- Eat small frequent meals.
- Eat snacks that are high in protein and calories.
- Eat whenever you are hungry.
- Rest before eating.
- Use attractive table setting.
- Eat with family and friends.
- Take medication for nausea one-half hour before meals.
- Avoid foods that cause bloating or gas, such as broccoli, cabbage, or beans.
- Avoid spicy foods.
- Nutritional supplements can be recommended by the dietician.
- Appetite-stimulating medications may be prescribed from physician.
- Eat sugar-free candy, gum, or mints.

NEUTROPENIC DIET

A neutropenic diet is for patients with weak immune systems. This diet will help protect the patient from bacteria found in some foods and drinks. The weakened immune system prevents your body from protection against these bacteria. Cooking foods well will help insure that all the bacteria are killed.

- Avoid fresh fruits and vegetables.
- Avoid raw or rare-cooked meat, fish, and eggs.
- Avoid freshly sliced deli meats.
- Avoid dry fruit and raw nuts.
- Avoid yogurt and yogurt products.

MOUTH SORES

Mouth sores are known as mucositis. This is a side effect that interferes with patient comfort as well as with intake of foods and liquids. Mucositis is an inflammation of the mucous membranes. This is a very common side effect of chemotherapy treatments.

- Assess gums and mucous membranes every eight hours or as needed.
- Routine mouth care before and after meals.
- Avoid mouthwashes that may contain alcohol.
- Drink through a straw to bypass sores and other tender areas of your mouth.
- Avoid flossing.
- Keep lips moist.
- Rinse mouth with a mixture of ½ teaspoon salt and 8 ounces of water every 1-2 hours during day and at least once during night.
- Do not wear dentures or dental plates while you sleep.
- Do not drink alcohol.
- Do not smoke.

PRECAUTIONS AT HOME

Chemotherapy precautions do not end when the patient is discharged home. They will continue even after the patient is discharged. Precautions are utilized to prevent the caregiver from an accidental chemotherapy exposure. The chemotherapy suppresses the immune system and the patient is still at risk.

- Flush toilet immediately after use. Flush toilet twice.
- Wash hands immediately with soap and water if urine, stool, or vomit comes in contact with your hands or other body parts.
- Caretakers should wear latex gloves if coming in contact with body fluids.
- If needed, wear disposable diapers or use disposable bed pads.
- Double bag waste before discarding in trash.
- Wash soiled linens as soon as possible and wash twice.
- Use condoms for approximately seven days to protect your partner from exposure to drugs in your body fluids.
- Women who are pregnant should avoid handling chemotherapy drugs or cleaning up waste products or spills.

DEFINITIONS

Antineoplastic agents	Drugs that slow down or prevent the growth of neoplasms
Chemotherapy	In the treatment of disease, the application of chemical reagents which are not harmful to the patient but which have a specific and toxic effect upon the disease-causing microorganism
Cytotoxic agents	Chemicals which destroy cells or prevent their multiplication. They are a group of compounds developed for use in cancer chemotherapy. The ideal agent for such use should destroy the fast-growing cancer cells without injuring the normal cells of the body
Extravasations	The escape of fluids into the surrounding tissues
Hazardous Drugs	Drugs that pose a potential health risk to health care workers who become exposed to them
Mucositis	Inflammation of the mucous membranes. Oral mucositis is a common side effect of some types of chemotherapy
Neoplasms	A new and abnormal formation of tissue, as a tumor or growth. It serves no useful function, but grows at the expense of the healthy organism
Neutropenic	Abnormally small number of neutrophil cells in the blood
Neutrophils	Our body's primary defense against bacterial infection
Phlebitis	Inflammation of the vein

ATTACHMENT "A"

Chemotherapy Drugs

A	Abraxane Accutane® Actinomycin-D Adria Adriamycin® Adrucil® Agrylin® Ala-Cort® Aldesleukin Alemtuzumab Alimta Alitretinoin Alkaban-AQ® Alkeran® All- transretinoic Acid Alpha Interferon Altretamine Amethopterin	Amifostine Aminoglutethimide Anagrelide Anandron® Anastrozole Anti-HER2 Monoclonal antibody Arabinosylcytosine Ara-C Aranesp® Aredia® Arimidex® Aromasin® Arranon® Arsenic Trioxide Asparaginase ATRA Avastin® Azactitidine
B	BCG BCNU Bevacizumab Bexarotene Bexxar® Bicalutamide BiCNU	Bischloronitrosourea Blenoxane® Bleomycin Bortezomib Busulfan Busulfex®
C	C225 Calcium Leucovorin Campath® Camptosar® Camptothecin-11 Capecitabine Carac TM Carboplatin Carmustine Carmustine Wafer Casodex® CBDCA CC-5013 CCNU CDDP CeeNU	Cerubidine® Cetuximab Chlorambucil Cisplatin Citrovorum Factor Cladribine Cortisone Cosmegen® CPT-11 Cyclophosphamide Cytadren® Cytarabine Cytarabine Liposomal Cytosar-U® Cytosine Arabinoside Cytoxan®

D	Dacarbazine Dacogen Dactinomycin Darbepoetin Alfa Dasatinib Daunomycin Daunorubicin Daunorubicin Hydrochloride Daunorubicin Liposomal DaunoXome® Decadron Decitabine Delta-Cortef® Deltasone® Denileukin diftitox DepoCyt®	Dexamethasone Dexamethasone Acetate Dexamethasone Sodium Phosphate Dexasone Dexrazoxane DHAD DIC Diodex Docetaxel Doxil® Doxorubicin Doxorubicin Liposomal Droxia TM DTIC DTIC-Dome® Duralone®
E	Efudex® Eligard TM Ellence TM Eliplen Eloxatin TM Elspar® Emcyt® Epirubicin Epogen Epoetin alfa Erbitux TM	Erlotinib Erwinia L-asparaginase Estramustine Estracyt® Ethyol Etopophos® Etoposide Etoposide Phosphate Eulexin® Evista® Exemestane
F	Fareston® Faslodex® Femara® Filgrastim Floxuridine Fludara® Fludarabine Fluoroplex®	Fluorouracil Fluorouracil (Cream) Fluoxymesterone Flutamide Folinic acid Folex® FUDR® Fulvestrant
G	G-CSF Gefitinib Gemcitabine Gemtuzumab Ozogamicin Gemzar® Gleevec TM Glidel ®wafer	GM-CSF Goserelin Acetate Granulocyte-Colony Stimulating Factor Granulocyte Macrophage Colony Stimulating Factor

H	Halotestin® Herceptin® Hexadrol Hexalen® Hexamethylmelamine HMM Hycamtin®	Hydrea® Hydrocort Acetate® Hydrocortisone Hydrocortisone Sodium Phosphate Hydrocortisone Sodium Succinate Hydrocortone Phosphate Hydroxyurea
I	Ibritumomab Ibritumomab Tiuxetan Idamycin® Idarubicin Ifex® IFN-alpha Ifosfamide IL-2 IL-11 Imatinib mesylate	Imidazole Carboxamide Interferon alfa Interferon alfa-2b Interleukin-2 Interleukin-11 Intron A® Iressa® Irinotecan Hydrochloride Isotretinoin
K	Ketoconazole	Kidrolase®
L	Lanacort® Lapatinib L-asparaginase LCR Lenalidomide Leucovorin Calcium Letrozole Leukeran Leukine TM Leukomax® Leuprolide Acetate	Leurocristine Leustatin TM Levamisole Liposomal Ara-C Liquid Pred® Lomustine L-PAM L-Sarcolysin Lupron® Lupron Depot® Lysodren®
M	Matulane® Maxidex Mechlorethamine Mechlorethamine Hydrochloride Medralone® Medrol® Megace® Megestrol Megestrol Acetate Melphalan Mercaptopurine Mesna Mesnex TM	Mexate® Meticorten® Mithracin® Mitomycin Mitomycin-C Mitotane Mitoxantrone M-Prednisol® MTC MTX Mustargen® Mustine Mutamycin®

	Methotrexate Sodium Methylprednisolone Mechlorethamine Hyd	Myleran® Mylocel TM Mylotarg®
N	Navelbine® Nelarabine Neosar® Neulasta TM Neumega® Neutrexin® Neupogen® Nexavar®	Nilandron® Nilutamide Nipent® Nitrogen Mustard Nizroal® Novaldex® Novantrone®
O	Octreotide Octreotide acetate Oncospar® Oncovin® Ontak® Onxal TM	Oprevelkin Orapred® Orasone® Ora-Testryl® Oxaliplatin
P	Paclitaxel Paclitaxel Protein-Bound Pamidronate Panitumumab Panretin® Paraplatin® Pediapred® PEG Interferon Pegaspargase Pegfilgrastim PEG-INTRON TM PEG-L-asparaginase PEMETREXED Pentostatin	Phenylalanine Mustard Plicamycin Platinol® Platinol-AQ® Platimun Prednisolone Prednisone Prelone® Procarbazine Procrit® Prokine® Proleukin® Prolifeprospan 20 with Carmustine Purinethol®
R	Raloxifene Recombinant Human interleukin-11 Revlimid® Rheumatres® RhIL-11	Rituxan® Rituximab Roferon-A® Rubex® Rubidomycin Hydrochloride
S	Sandostatin® Sandostatin LAR® Sargramostim Soltamox	SPRYCEL TM STI-571 Streptozocin SU11248

	Solu-Cortef® Solu-Medrol® Sorafenib	Sunitinib Sutent®
T	Tabloid® Tamoxifen Tarceva® Targretin® Taxol® Taxotere® Temodar® Temozolomide Teniposide TESPA Thalidomide Thalomid® TheraCys® Thioguanine Thioguanine Tabloid®	Thiophosphoamide Thioplex® Thiotepa TICE® Toposar® Topotecan Hydrochloride Toremifene Citrate Tositumomab Trastuzumab Tretinoin Trexall TM Trimetrexate Gluconate Trisenox® TSPA TYKERB®
V	Vasanoid® VCR Vectibix TM Velban® Velbe® Velcade® VePesid® Vesanoid® Viadur TM Vidaza® Vinblastine	Vinblastine Sulfate Vincasar Pfs® Vincristine Vinorelbine Vinorelbine tartrate VLB VM-26 Vorinostat VP-16 Vumon®
W	Wellcovorin®	WR 2721
X	Xeloda®	
Y	Yttrium-90	
Z	Zanosar® Zevalin TM Zinecard® Zoladex®	Zoledronic acid Zolinza Zometa®

2	2-CdA 2-Chlorodeoxyadenosine	2-deoxycoformycin
5	5-Azacitidine 5-FU	5-Fluorouracil
6	6-Mercaptopurine 6-MP	6-TG 6-Thioguanine
13	13-cis-Retinoic Acid	

(PDR, 2006)

ATTACHMENT "B"

INTRAVENOUS CHEMOTHERAPY FLOWSHEET

ADDRESSOGRAPH

ALLERGIES: Penicillin, E-Mycin, Eggs

NEEDLE DEVICE CODES	SITE CONDITION	LOCATION CODES	
IVC- Intravenous Catheter CVP- Central Line ML- Mid Line PIC- Peripherally Inserted Catheter	1. Normal 2. Tenderness 3. Palpable Cord 4. Warm to Touch 5. Erythema 6. Purulent Drainage 7. Other	H- Hand F- Forearm UA- Upper Arm J- Jugular R- Right O- Other	W- Wrist AC- Antecubital SC- Subclavian FA- Femoral Artery L- Left

Tubing Changes

Date/Time	Initials
4/5/07 8am	cla

Dressing Changes

Date/Time	Initials
4/5/07 7:45 am	LG

DATE	TIME	DEVICE/ LOCATION	SITE CONDITION	SOLUTION	RATE	AMOUNT REMAINING	AMOUNT ABSORBED	COMMENTS	INITIALS	VERIFYING INITIALS
4/5/07	8a	R Pic	1	5 FU	43	1048	0	New bag	cla	LG
4/5/07	905a	R Pic	1	5 FU	43	1004	44		LG	
4/5/07	955a	R Pic	1	5 FU	43	962	42		cla	
4/5/07	11a	R Pic	1	5 FU	43	919	43	Blood return noted	cla	

Chemotherapy Precautions
Test Your Knowledge
Post-test

Directions: Select the correct answer for each of the following questions.

1. Using the safety guidelines for chemotherapy administration, what is the designated chemotherapy excretion safety period?

_____48 hours for all drugs
_____72 hours for all drugs
_____7 days for all drugs
_____72 hours or 7 days, depending on the drug

2. What agency has established guidelines for the handling and administration of chemotherapy agents for the health care professional?

_____Joint Commission on Accreditation of Healthcare Organizations (JACHO)
_____American Cancer Society (ACS)
_____National Cancer Institute (NCI)
_____Occupational Safety and Health Administration (OSHA)

3. When handling the waste of individuals receiving chemotherapy, adequate protection includes

_____gloves.
_____gloves, gown, face protection.
_____gloves, gown.
_____gloves, face protection.

4. Wearing two pairs of chemotherapy gloves when a gown is worn serves to protect the skin from contamination while the gown is being removed. Listed below are the steps to be followed when applying the two pairs of chemotherapy gloves and the gown. Number the steps from 1 to 3 to correspond with the correct order in which you would perform these activities:

_____apply the second pair of gloves
_____apply the first pair of gloves
_____apply the gown

5. When transporting a patient who is receiving chemotherapy to the radiology department, the intravenous catheter becomes dislodged and the chemotherapy agent spills onto the floor. What action would you take?

_____ Take the patient back to his room and call housekeeping to clean the spill
_____ Secure the area so no one is accidentally exposed before the spill can be cleaned up and follow the directions on the spill kit for cleanup
_____ Reinsert the intravenous catheter, call housekeeping to clean up the spill, and continue with the patient to the radiology department
_____ Call the nurse's aide assigned to that team to clean up the spill

6. The contents of the Chemotherapy Drug Spill Kit includes all but

_____ shoe covering.
_____ surgical mask.
_____ chemo waste bags.
_____ a scooper with detachable scraper.

7. Your patient has developed small mouth sores after receiving his chemotherapy. Teaching should include

_____ do not drink through a straw as the straw puts pressure on the ulcers causing more pain.
_____ any type of over-the-counter mouthwash will be effective in reducing the sores.
_____ avoid hard candy or gum if your mouth or throat is sore.
_____ avoid citrus fruits, tomatoes, and alcoholic beverages.

True or False:

8. Yellow waste bags are left in the patient's room when a patient is receiving chemotherapy until either the chemotherapy is finished or the bag becomes full.

_____ True   _____ False

9. A chemotherapy trained registered nurse must monitor the chemotherapy infusion.

_____ True   _____ False

10. A significant chemotherapy spill is 5 milliliters or larger.

_____ True   _____ False

Answers to Pre/Post Tests

1. 72 hours or 7 days, depending on the drug (page 11)

2. Occupational Safety and Health Administration (OSHA) (page 3)

3. gloves, gown, face protection (page 8)

4. 3-1-2 apply the first pair of gloves, apply the gown, apply the second pair of gloves (page 9)

5. Secure the area so no one is accidentally exposed before the spill can be cleaned up and follow the directions on the spill kit for cleanup (page 15)

6. surgical mask (page 14)

7. do not drink through a straw as the straw puts pressure on the ulcers causing more pain (page 28)

8. False (page 10)

9. False (page 3)

10. True (page 13)

ACKNOWLEDGEMENTS

I would like to thank everyone who made this booklet possible. If I fail to acknowledge you, it is not intentional, but merely an oversight.

First and foremost, I would like to thank my family. I would like to thank my daughter, Bethany, for her patience. She spent many hours proof reading my papers and teaching me computer skills. She gave me encouragement as I was continuing my education. She also gave me my grandson, Lucas, who helped keep me awake at night so that I could write papers and study. I would like to thank my sisters and brothers for their encouragement and support. I would also like to thank my Mother. Even though she is no longer with me, she was a big influence in my professional career.

Thank you to the staff of 3 Mansfield, UPMC McKeesport for helping to provide some of the pictures used in the booklet. You are a great group of people to work with and I thank you. Many thanks to the staff of Hillman Cancer Center, Pittsburgh, Pennsylvania who provided me with the pictures of intravenous complications. Thanks to Danielle Spirnak, who gave her time to mentor me during my practicum.

Finally, I would like to thank two of my professors. Dr. Mary O'Connor from California University of Pennsylvania for believing in me and Dr. Theresa Shellenbarger from Indiana University of Pennsylvania for all of your encouragement, guidance, and assistance in completing this booklet.

REFERENCES

American Cancer Society. (2006). *What are the different types of Chemotherapy drugs?*

    Retrieved May 25, 2007, from http://www.cancer.org

Kendall, LTP. (2005). *Products*. Retrieved March 31, 2007, from

    http://www.Kendall-LTP

PDR Physicians Desk Reference (2006). *Nurse's Drug Handbook.* Clifton Park, NY:

    Thomson.

Occupational Safety and Health Administration. (2006). *Technical Manual, Section VI,*

    Chapter 2, Retrieved February 23 2007 from

    http://www.osha.gov/dts/osta/otm/otm_vi/otm_2.html

Polovich, M. (2004). *Safe handling of hazardous drugs*. Retrieved February 16, 2007,

    from http://www.nursingworld.org

Rosenthal, K. (2006). *Infiltration: an ounce of prevention*. Retrieved March 29, 2007,

    from http://www.fswp.org/employment

www.ingramcontent.com/pod-product-compliance
Lightning Source LLC
Chambersburg PA
CBHW051054180526
45172CB00002B/634